Deconstructing Modern Physics

Greg Feild

January 28, 2022

About the author

I earned a Ph.D in experimental high energy physics from the Pennsylvania State University working on HERA at DESY in Hamburg, Germany studying photoproduction and deep inelastic scattering in electron-proton collisions.

I did my postdoctoral studies with Yale University working at Fermilab on the CDF experiment at the Tevatron. My primary research interest was particle hadronization in quarkonium production in proton-antiproton collisions.

I have always found the standard model unsatisfactory, even when I had less of an understanding of it than I suppose myself now to have! These misgivings about the standard model are quite distinct and separate from my disdain for the Copenhagen interpretation of quantum mechanics. Putting all mathematical complexity aside, these theories are conceptually incoherent and rationally inexplicable. To claim this is just how the world works is merely begging the question and no answer to skeptical inquiries.

Abstract

The field of modern theoretical physics is in need of a major course correction. In this book, we argue that current physical models and physics theories have way too many unobservables, way too many forces, way too many fields, and just way too many variables to function as a proper theory at all.

Forget about revolutions and paradigm shifts. Revolutions are never really more than surface ripples. To save physics, we need to keep shoring up the original foundation while removing the hideous, gilded, baroque facade it has acquired over the last one hundred years. Thus, we offer a new metaphor and a new mechanically motivated model for a new millennium.

an ahistorical approach

Introduction

Should we even speak of modern physics anymore? Most of the maladies of modern physics have been inherited from classical theory, where they were similarly problematic. I am thinking in particular about field theory, of course, and the cascading and ever more far-fetched and abstruse abstractions it has spawned.

Do we really share the astonishment of the physicists of the turn of the last century at the mysterious aspects of the behavior of elementary particles, or is such an attitude merely (sincerely) feigned? Our astonishment should be that such talk still abounds. Is such an attitude still constructive for doing physics?

Can we reign in current theories and speculators? Can we still accept explanations that are really no explanations at all? Can we not with hindsight, and foresight, begin again and build a better physical and metaphysical model of particles and their interactions? I'd like to think that we can and that we are well underway.

My main gripe with *all* of physics (classical and quantum mechanical; general relativity, solid state physics, etc.) is the *reification of fields.* People believe, or must operate as if they believe, that fields are real actual things (items, objects, waves?) permeating all time and space and responsible for the behavior of the microcosm as well as the macrocosm. Some people are quite serious in this belief.

As everyone knows, fields are a convenient calculational and experimental device, derived from the original laws governing the forces between two massive or charged bodies. Once these fields became more than just mathematical placeholders for the force that a test charge would feel at a particular point in space and time, it became requisite that these fields store energy, suffer stress and strain, etc., and worst of all, particles began interacting with themselves!

Particles do not interact with themselves. <u>Fields were not meant for such close up work</u>.

At the next level of abstraction from the (vector) field we have the scalar potential field; arguably much more useful for many physical problems and perhaps more beautiful, but more dangerous as well. Are we not already on the slippery gradient? From the scalar and vector potentials, we then can construct the four-potential; a concept we quite like, actually, and which we have reappropriated for the foundation of our revised theory. For us it is an inflexion point between the two extremes of fields and gauges. Gauge transformations, like fields, are mathematical devices designed to make certain physical calculations or mathematical representations easier or more compact, etc. From these humble beginnings they have become generators of all sorts of charges and forces. People like to say Nature “knows” about gauge invariance !

Nature does not know about gauge invariance. Nature doesn't know about groups.

Nature does not know about symmetry; hidden, broken, or otherwise. The only symmetries in our model are due to translational and rotational invariances. And parity reflection, probably.

I believe all symmetries uncovered by the standard model are mere inventions; mistakes due to invalid fundamental assumptions, theories constructed to keep these assumptions alive. Nature *is* symmetric in our theory. How could it be otherwise? For every action (appropriately defined) there is an equal and opposite reaction. If there is a spin to the left, there is a spin to the right, etc. Current notions about parity violation stem from misunderstandings about the nature of particle spin, helicity, and how to *assign a parity* to a particular particle. I believe we have shown this sufficiently in previous papers, but shall discuss some of these issues more in this book.

The nature of spin is the keystone of our new model. In our interpretation of quantum mechanics, particles don't spin, or "have" spin; particles *are* spin. We have proposed a mechanical model of spin, that is not quite 'classical' (we're dealing with spinors, after all), but one which appeals to both the human imagination and intuition, and is useful for explaining physical phenomena both qualitatively, and quantitatively as well. From the spin we can derive the particle mass, energy, angular momentum, and most likely the charge.

In our model, an electron, for example, has three linearly independent spin components; $s_x = s_y = s_z = h^{bar}/2$. The direction and orientation of these three spin components are fixed and constant and continuous and are maintained during a particle's motion discounting interactions with other particles, of course. In principle (and perhaps only ideally), we can determine two of these three components of a particle's spin; one from the particle helicity (the projection of the spin s_x along the direction of travel), and one from the particle's behavior in a magnetic field (the projection of the spin s_z 'up or down'). This then leaves the spin component s_y free to point 'up or down' <u>according to some *initial conditions*.</u> This spin orientation is unknown only to us, hence the appearance of complex numbers in the spin matrix σ_y. The spin system of equations is mathematically *underdetermined* but *the physical system is completely determined* (and known 'to itself' !).

Electrons are 'matter' and spin to the left. Due to the spinor nature of the electron it *always spins to the left,* no matter whether it is pointing up or down. Its helicity is always left-handed. Positrons spin to the right. When an electron and a positron meet or collide, they combine to form a virtual photon in a completely 'natural and physical way'. There is no magical annihilation or burst of energy, no vacuum, no conjuring of photons. Only particle combinations and decays.

In our model, *this spinning to the left is negative charge*. As an electron is accelerated, its angular momentum or rate of spin to the left increases, and technically one could say that its electric charge increases, as is well known. The reason why the electron has a fixed 'rest charge' is that there is a fundamental lower limit on the frequency of the electron 'spin'.

The origin and mechanical nature of the lower bound on this rate of spin is one of the mysteries of our model. The next mystery, of course, is what fixes the lower frequencies of the muon and tau in such a manner that their rest charges are the same as that of the electron.

The neutrino differs from the electron only in that there is no lower bound on the frequency of its rate of spin. Hence, the neutrino is a rest-massless matter particle. The photon is a rest-massless mass particle. The difference between mass and 'matter' is the difference in the spin projections and modes of propagation. We have discussed this previously and shall not pursue it further here.

Now is as good a time as any to introduce the Metaphysic we have developed for our model over the last several books. Each particle, or system of particles, seeks to be in their ground state, or lowest possible energy state given the circumstances. (One might say all particles want to be 'at rest'.) This is essentially the same idea underlying current classical and modern physics with the minimization of the action given an appropriate Langrangian. In our new model, we endorse a similar approach, but with the minimization of work. We try to avoid the concept of potential energy (and forces, if possible), focusing instead on the kinetic energy of the particle or system of particles under consideration.

But, back to our Metaphysic. Each particle is connected to every other particle via a virtual photon, hence our retention of the four-potential from the standard model. However, this four-vector represents the energy and momentum, or 'wave function', of the virtual photon connecting the two particles, shuttling energy, momentum, and angular momentum back and forth. Each particle considers every other particle as a possible sink for its excess energy, and vice versa, of course! Two electrons spinning (by definition) in the same direction will each try to dump their excess angular momentum or rotational kinetic energy to the other. This constant contest or perpetual exchange will accelerate or drive the electrons apart due to conservation of energy and angular momentum and the explicit nature of our new model of electron propagation. In the case of an electron and a positron, the mutual exchange of opposite angular momenta will cause the two particles to move toward one another. (We shall discuss gravity in a moment.)

Consider a two particle universe consisting of an electron and a positron for purposes of conservation! The two particles interact via one virtual photon. There is no need to fill the universe with fields or infinite amounts of energy.

The upshot of this discussion is that *angular momentum* or *mass* is the fundamental *universal charge*. Electric charge makes its appearance as a fundamental, irreducible, constant spinning to the left; the rate dictated by the rest mass of the electron.

Clumps of neutral matter, separated at a distance, no longer transfer any net angular momentum via the two 'transverse' channels of the virtual photon leaving only the spacelike or longitudinal channels for net momentum transfer, and these two channels are inherently attractive. This is our theory of gravity.

It must be remembered that mass is derived from and/or is reducible to angular momentum. This includes intrinsic rotational angular momentum as well as relative angular momentum. So the gravitational force between two massive bodies is dictated by the total mass-energy difference, or relative angular momentum between the two bodies, and this includes the rotation of each body and the *relative motion* between the two bodies. Hence our expression of the gravitational force is a direct analog of the equation for the electromagnetic Lorentz force. In fact, in our model, the gravitational and electromagnetic forces are completely equivalent in their behavior and mathematical expression; only the strengths of the coupling differ.

So, rather than trying to unite quantum mechanics and general relativity, we decided general relativity had to go. In our model all gravitational interaction is mediated by virtual photons. *All* interaction is mediated by virtual photons. There are no massive gauge bosons. There are no leptonic propagators; hence in our model the photon must be a self-coupling gauge boson. This is necessary to account for the Compton effect as well as the bending of starlight by the sun!

In addition, there are only two forces; gravity and electromagnetism. The weak and strong forces represent the manifestation of the two extremes of *relativistic gravity* at the subatomic level. Call it Quantum Gravity if you must! There are no quarks or gluons, no gravitons, no Higgs boson. There are only leptons and photons.

This is our new model in a nutshell. All of these ideas have been presented to varying degrees in previous publications. It would be 'impossible' and redundant to reproduce all the arguments here. However, we would like to look at the fundamental nature of elementary particles and their modes of propagation one last time since this is the bedrock of our theory.

After a brief overview of the nature of elementary particles and how they interact (in our model), we will give a report on the status of two important aspects of our model that need more development, namely; the phenomenon of quantum mechanical electromotive force and the origin and nature of the particle families. Finally, we shall make some cautious statements about time and space.

Elementary particles

As stated in the introduction, in our model an electron (or any lepton) has three linearly independent spin components; $s_x = s_y = s_z = h^{bar}/2$. An electron *is* the vector $L = \sqrt{3}/2\ h^{bar}$. This vector 'precesses' (in a special spinor way) at a minimum rate determined by the electron rest mass. As the electron propagates and precesses, the projection of the angular momentum along the direction of travel varies sinusoidally (although the polarization remains constant.)

Our model of electron spin and mode of propagation is depicted in Figure 1.

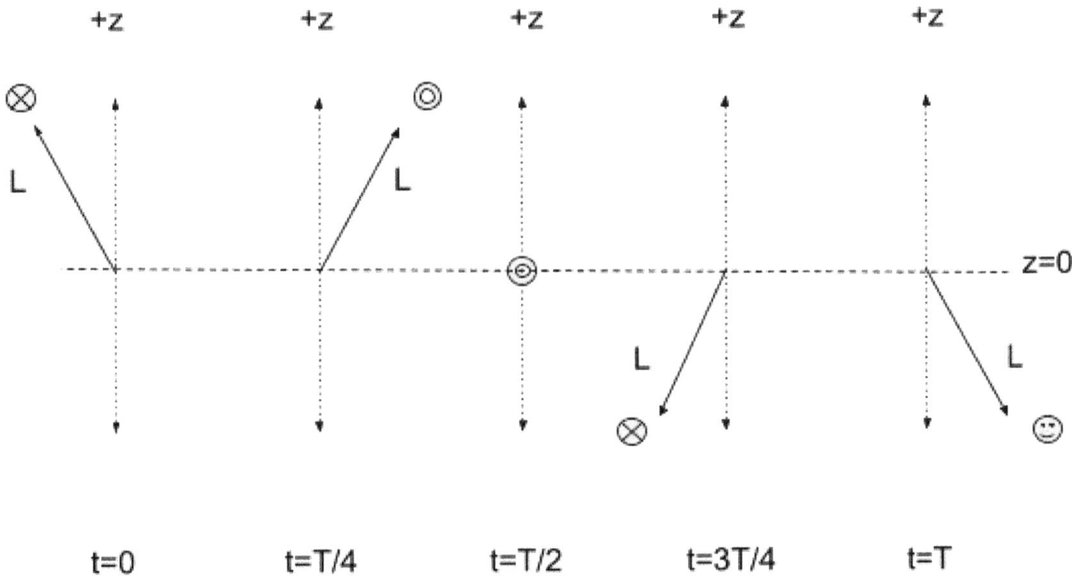

Figure 1: At rest with a lepton traveling in the z-direction. The spin angular momentum vector 'precesses' about the direction of motion, tracing out a closed, three dimensional figure eight. The x symbol represents motion into the page. The dot symbol represents motion out of the page. At time T/2, we see the angular momentum is *perpendicular* to the direction of travel. To represent an antilepton, simply swap the x symbols and the dot symbols.

Referring to Figure 1, we see that at time T/2 there is no projection of the angular momentum along the direction of travel, hence an electron is unable to absorb or *interact with a real photon at these sorts of nodes*. These nodes are where the particle catches up with its virtual interactions and/or bounces off a container wall!

We see from Figure 1 that the angular momentum vector 'circles' the direction of propagation twice before returning to its original position. Thus, rather than a point particle, the electron is a special simple harmonic oscillator, or a spinor.

The electron has an operational, or effective, radius corresponding to its wavelength. Even at rest an electron has a wavelength, the Compton wavelength, since an electron is always spinning or 'precessing' about some preferred axis or direction. In our model we want to express everything in terms of rotation, so we have recast, derived, or reimagined the mass and energy of the electron as or in terms of moments of inertia and angular frequencies. A summary is presented below.

In our model the angular momentum and the energy of an electron can be written

$$L = \sqrt{3}h^{bar}/2 + I\omega \tag{1}$$

$$E = h^{bar}\omega_0 + I\omega^2 \tag{2}$$

$$\omega_0 = 2\pi c/\lambda_0 \tag{3}$$

where λ_0 is the Compton wavelength of the electron, which we also take to be the 'rest radius'.

We define the velocity dependent moment of inertia $I(\lambda)$ to be

$$I = m\lambda^2/(2\pi)^2 \tag{4}$$

From equations (1) and (2) one can see that the 'rest angular momentum' and 'rest rotational kinetic energy' are given by

$$I_0\omega_0 = \sqrt{3/2}\, h^{bar} \tag{5}$$

$$h^{bar}\omega_0 = m_0c^2 = I_0\omega_0^2 \tag{6}$$

In our new model, the magnetic moment of the electron is thus velocity dependent (6,13,14,18);

$$\mu = (e/m_e)(h^{bar}/2c)(1 + \tfrac{1}{2} v^2/c^2 + \tfrac{3}{8} v^4/c^4 + \ldots) \tag{7}$$

And we suggest the gravitational magnetic moment of the neutrino is

$$\mu = (h^{bar}/2c)(1 + \tfrac{1}{2} v^2/c^2 + \tfrac{3}{8} v^4/c^4 + \ldots) \tag{8}$$

In our model the proton consists of two positrons and one electron bound by an interplay of the electromagnetic and relativistic gravitational forces. One can easily imagine the two positrons (one spin up and one spin down) forming a closed orbital shell around the electron nucleus.

A hydrogen atom would then contain *two electrons and two positrons*. There is no missing antimatter and no matter-antimatter imbalance in our model.

Here is a summary of the coupling strengths for the 'four forces' in our model.

The electromagnetic coupling 'constant', alpha, is (14);

$$\alpha = \alpha_0(1 + (v/c)^2 + (v/c)^4 + \dots) \quad (9)$$

$$\alpha_0 = e^2/4\pi\varepsilon h^{bar}c \quad (10)$$

The gravitational coupling constant is

$$\alpha_G = (m_e^2 G)/(h^{bar}c)\ (1 + (v/c)^2 + (v/c)^4 + \dots) \quad (11)$$

The weak coupling constant is

$$\alpha_W = (m_\nu^2 G)/(h^{bar}c)\ (1 + (v/c)^2 + (v/c)^4 + \dots) \quad (12)$$

And finally, the strong coupling constant (14) is

$$\alpha_S = (G/4\pi\varepsilon)^{1/2}\ (2m_e e/h^{bar}c)\ (1 + \tfrac{1}{2} v^2/c^2 + \tfrac{3}{8} v^4/c^4 + \dots) \quad (13)$$

In the universal model, all the 'constants' run, because *the fundamental coupling charge* of a particle *is the relativistic mass-energy* of the particle.

The choice of the rest mass particles appearing in the coupling constants is conventional. In our model the neutrino no longer even has a rest mass. Perhaps some practical lower limit could be imposed.

In our model for two body interactions, only the 'lowest order diagrams', and only those involving 'one photon exchange' would contribute to the matrix elements. All higher order corrections would be due to the expansions of the particle masses in terms of v/c. Otherwise, the calculation of scattering cross sections, etc., would proceed much as they do now.

Will this approach yield sensible results agreeing with experiments? I don't know !

In our model the formulation of the Dirac equation is

$$H \psi = c\alpha \bullet p \psi \qquad (14)$$

$$i\hbar \partial\psi/\partial t = -ic\hbar\alpha \bullet \nabla\psi \qquad (15)$$

and is satisfied by, or solved with, the universal wave function

$$\psi = \exp(i(p \bullet x - (m-m_0)c^2 t)/\hbar) \qquad (16)$$

We can see this wave function contains the rest mass (or the global charge) and the relativistic mass (or the local charge). This is what one might call ‘gauge invariance’. Certainly under Lorentz transformations the rest mass is constant and the relativistic mass is, well, relative.

This universal wave function describes the behavior of a spinor as depicted in Figure 1. This spinor satisfies the Dirac equation with two *positive energy solutions spinning to the left* and two *positive energy solutions spinning to the right.* We have real particles doing real things that we can at least conceptually picture and understand using mechanical models based on experience.

There are no particles traveling backwards in time, no ‘holes’ in the electron sea, etc. We have to pause and wonder at the origin of such absurd beliefs, why they were so readily adopted, and why people still fundamentally believe the same thing today, although in a more supposedly “sophisticated” way, of course. There seems to be no foundation for them, no ontological or epistemological basis or justification for making such explicit declarations. One cannot appeal to Pragmatism to support such grand Metaphysical claims.

Another feature of our formulation of the wave function is that it removes the explicit rest mass term from the Dirac equation; a term which has caused *all sorts of problems* for the standard model as far as I can tell. The ultimate nonsensical, paradoxical result is that any ‘complete’ theory should somehow account for the ‘origin’ of particle mass. (It seems that the idea of a massive particle is mysterious and problematic, while the notion of interacting fields is not.)

What is puzzling is how this slow sedimentation of ad hoc metaphysical assumptions became assimilated into physical lore and our laws of nature over the years. Young people entering physics today are already prejudiced and influenced by the inane nattering of the “Sunday scientists” and are in no position to critically examine the foundations of their own discipline.

To add an interaction term to our Dirac equation, we do not make the classical canonical electromagnetic replacement for the momentum, but simply include the term ieA. The four vector A represents the energy and momentum of the virtual photon or photons connecting the particle to the other particles involved.

Particles do not move through pre existing fields, no matter how attractive or convenient this picture may be. Particles, even test charges and experimental probes, are constantly interacting with everything and carry or drag these vertices around, constantly responding to and simultaneously altering them. Fields are a wonderful calculational device, a beautiful concept, but they are not real. They have no properties of their own.

Finally, let us consider our Dirac equation for a particle at rest. As we know, there is an uncertainty in the position and momentum, even for a particle 'at rest'. This conception is for a *point particle.* In our model, the electron is not a point particle, but a momentum vector tracing out a fundamental volume with a radius we characterize by the Compton wavelength. For a particle at rest, $\Delta x = \Delta y = 0$, while $\Delta z = \lambda_0 = h/m_0c$. Equation (14) becomes

$$H\,\psi = c\alpha \bullet \Delta p\,\psi \tag{17}$$

$$\Delta p = h/\Delta z = h/\lambda_0 = m_0 c \tag{18}$$

$$H\,\psi = \alpha \bullet m_0 c^2\,\psi \qquad ; \quad \alpha = \alpha_z \tag{19}$$

Even though we have removed the rest mass explicitly from the Dirac equation it still applies to a particle at rest as it seems it should. The rest mass of the electron is due solely to the rotational kinetic energy of the 'spinoring' of the angular momentum vector $L = \sqrt{3}/2\ h^{bar}$. The (sinusoidal) projection of the angular momentum of the electron along the z-axis is h/2. This allows the electron to absorb and emit energetic photons of angular momentum L = h.

We have proposed that the electron always spins to the left regardless of whether it is pointing up or down. Referring to Figure 1, we can see that the only conceivable difference between spin up and spin down would be a phase difference of $\pi/2$ in the relative rotation between any two angular momentum vectors. Nothing can point up all by itself.

Quantum mechanical electromagnetic induction

In our model, a mass current produces a gravitational magnetic field analogous to the magnetic field produced by a charge current. The original motivation for this perhaps not very original idea was to give the neutrino a magnetic moment of one sort or another. This proved fruitful in putting gravitation and electromagnetism on an equal footing and paved the way for our suppositions concerning the microscopic and macroscopic behavior of gravity.

(As an aside, we are no longer trying to force the neutrino and electron magnetic moments to have the 'same units', realizing that appropriate factors of α_G and α, or their roots, will solve the problem of the difference of charge between the two in any concrete calculations.)

Any effects of macroscopic mass currents are (usually) negligible in that they carry no net intrinsic angular momentum or spin. We remind the reader, again, it is the preferential spinning to the left or to the right of electrons and protons that make them stand out as electric charge.

In our original conception of the nature of electric charge, we assumed that there were certain 'resonances' in the frequency of spinning mass which were fixed, or quantized. The spinning electron mass induces a magnetic field which then induces an electric field which we experience as the electron magnetic moment and electric charge. (The muon mass would indicate the next resonance in this phenomenon of quantum mechanical electromagnetic induction, then the tau, and then the continuum produced by electric currents that we exploit in transformers and so on!)

Now, it is clear that the mass, or charge, and magnetic moment of a particle are two manifestations of one and the same phenomenon, namely spin angular momentum. It is just a matter of considering the different moments of mass. From afar, the particle appears as a point charge since the local oscillations of the angular momentum of the source charge have little effect at long distances. It is only 'up close' that this spinning becomes important as an effective time varying periodic electric charge.

So, an electron is not a spinning electric charge producing a magnetic field, but a spinning angular momentum that manifests as both. Neither electricity or magnetism are fundamental, as we know, but reflect a simpler underlying mechanism. You can't have one without the other.

The electric charge of the electron is fixed because the rest mass of the electron is fixed. The rest mass of the electron is fixed because it is! If the electron didn't have a fixed rest mass we wouldn't be here. One cannot argue with logic like that.

Particle families

In our formulation, electric charge is not an extra attribute of a particle above and beyond the angular momentum or mass of the particle. In fact, the electric charge can be considered strictly as a coupling constant denoting any particle with a *fixed minimum angular momentum*. Thus, it is not a mystery why the muon and the tau have the same charge as the electron, but only why their rest masses are what they are. The electron, muon, and tau *neutrinos* differ only in their initial lower frequency bounds, dictated by the charged particles' rest masses.

All three leptons are essentially the angular momentum vector $L = \sqrt{3}/2\ h^{bar}$. The three leptons only differ in the minimum rate of precession of this angular momentum vector. In keeping with our theme of the minimization of work, or the minimization of changes in kinetic energy during any interaction, as the foundational principle of physics, we decided that all interactions must minimize the difference between the rest masses and the kinetic energy of the particles involved, thus creating the need for families. Crudely put, at some point it is more expedient for nature to produce a slow muon rather than a fast electron.

Our model of muon decay is presented in Figure 2. The muon sheds its excess matter in the form of neutrinos. The 'massless' virtual electron is really just a glorified virtual photon since it *is no longer required to carry charge*. This is a new tweak to our model.
There is a factor of $(\alpha_W)^{1/2}$ at each vertex.

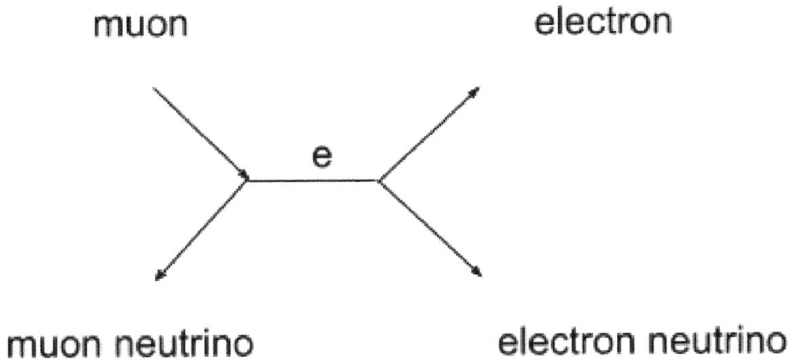

FIgure 2: Muon decay. The 'propagator' is virtually a virtual photon!

The leptonic table:

LEPTONS			ANTI-LEPTONS	
electron	electron neutrino	PARITY ⇔	electron antineutrino	positron
⇐	CHARGE	MASS ⇅	CHARGE	⇒
muon	muon neutrino	PARITY ⇔	muon antineutrino	anti-muon
⇐	CHARGE	MASS ⇅	CHARGE	⇒
tau	tau neutrino	PARITY ⇔	tau antineutrino	anti-tau
⇔	mass isospin	charge isospin ⇳	mass isospin	⇔

TABLE 1: The leptons and their interrelations; or the kleptogenesis of the leptoquarks.

Any lepton can be 'generated' from any other by the appropriate applications of the parity operator, the mass isospin operator, and our newly proposed 'charge isospin' operator. Mass isospin is the same as weak isospin in the standard model. The idea, of course, is that the neutrino and the electron have the 'same' mass and only differ in electric charge, and that the electron, muon, and tau have the same electric charge but only differ in mass. Particles and antiparticles differ only in parity; i.e. spin to the left or right !

Using various combinations of the step up and step down operators of SU(2) and SU(3), plus the 'parity operator', we can write any quantum mechanical interaction current in terms of the 'fundamental' neutrino neutral current.

Maybe nature knows about groups.

Time and space

In the beginning there was the universe. A universe much as we see now, with a similar distribution of mass and matter, stars and galaxies, intergalactic gas clouds and black holes. An eternal, boundless, flat and static, albeit self-renewing, universe where entropy need not apply.

One might say in the past, we've been "all over the map" concerning time and space. Perhaps. But we must remember we are dealing with essentially metaphysical concepts and trying to appropriate and quantify them to function in our physical theories and equations. In this sense the definitions of time and space are no different than the definitions of mass, force, and acceleration, except for the fact, of course, that acceleration assumes or implies time and space. Nothing can be fixed, or isolated, or defined without reference to something else.

This is essentially our theory of time and space. You need at least two objects or particles to define a space, plus some relative motion between the two to fix a velocity. A single particle plus an observer really won't do! Once you fix your observer "by definition" at the origin of your inertial coordinate system, it is only the relative separations, velocities, and accelerations of the particles that matter for physics. In any fixed coordinate system, time and space really do act as 'containers'; that is, *they do not act at all*. The space coordinates of any inertial coordinate system are three independent variables, while the *time* is merely a counter, dependent on some *relative motion in the coordinate system* thus defined. Time is a parameter, not a variable.

white space

Conclusion

We have no problem with the mathematics of quantum mechanics. Quantum mechanics fits quite nicely into and is in fact the foundation of our model. What we take issue with, and must vehemently protest, is the magical and mystical interpretations of quantum mechanics that have essentially become gospel for most people. Our mechanical model of the spinor accounts for the double valued nature of the electron wave function, the Heisenberg uncertainty principle, the Born rule; and qualitatively at least, the double slit experiment and quantum mechanical tunneling. All this is fairly clear in the case of a single free particle, at least. I believe the case for tunneling could be given a mathematical footing fairly easily.

In our model particles always have concrete values of spin angular momentum; both in magnitude and orientation. There is no spooky action at a distance. There is no wave function collapse. When an experimenter sets up an experiment, they do not create a wave function which they subsequently collapse. The idea is absurd. The wave function is a theoretical device used to calculate experimental outcomes. In our reconfiguration, the wave function is not a probability wave but rather represents the real time evolution or status of a particle's (or system's) energy, momentum, and angular momentum as the particle propagates or 'evolves' under the influence of mutual, equal and opposite, forces.

Our 'Equivalence Principle' is that everything is angular momentum; inertial mass, gravitational mass, electric charge, leptons, photons and … that's all there is! The question of particles possibly possessing two different types of mass always seemed to me more like a posturing than a serious scientific concern. What could this even mean? Perhaps it is a serious consideration for people eventually destined to also assign the lowly particle flavors, colors, and weak charge in addition to mass and electric charge!

In an attempt to 'mathematically' describe particle creation and destruction the metaphorical idea of the quantum mechanical vacuum spawned a metaphysical nightmare. The vacuum became more pervasive and more replete with convenient properties, even as it became ever more real in people's minds the more problems it was able to solve or resolve. A bubbling, magical cauldron indeed !

Particles are not created and destroyed. Particles merge, or combine, and then decay appropriately (i.e., in the most 'efficient' way) given the energetically possible available particles and the final energy-momentum phase space, and so on. If this process is best represented mathematically with creation and destruction operators then so be it, but *don't reify the vacuum*.

Nature abhors the quantum mechanical vacuum.

Books by Greg Feild

1. “A quantum mechanical theory of gravitational interactions”
   CreateSpace Independent Publishing, 8/29/2016

2. “Observations on the quantum mechanical nature of gravity”
   CreateSpace Independent Publishing, 10/8/2016

3. “On gravitation and electric charge”
   CreateSpace Independent Publishing, 10/29/2016

4. “On spin, mass, and charge”
   CreateSpace Independent Publishing, 11/29/2016

5. “On angular momentum, acceleration, and absolute motion”
   CreateSpace Independent Publishing, 1/1/2017

6. ”The Sinister Universe”
   CreateSpace Independent Publishing, 3/1/2017

7. ”On Parity and Isospin”
   CreateSpace Independent Publishing, 4/11/2017

8. “Reflections on the Sinister Universe”
   CreateSpace Independent Publishing, 5/12/2017

9. “On Current Physics”
   CreateSpace Independent Publishing, 6/11/2017

10. “A Critical Examination of Classical and Quantum Mechanical Waves”
    CreateSpace Independent Publishing, 6/18/2017

11. “On wave particle duality and the quantum of action”
    CreateSpace Independent Publishing, 7/6/2017

12. “On matter, mass, and motion”
    CreateSpace Independent Publishing, 9/14/2017

13. “On action and reaction”
    CreateSpace Independent Publishing, 9/24/2017

14. “A quantum mechanical theory of everything”
    CreateSpace Independent Publishing, 11/5/2017

15. “On Interaction”
CreateSpace Independent Publishing, 4/21/2018

16. “On Rotation”
CreateSpace Independent Publishing 8/19/2018

17. “Revenge of the Sinister Universe: The Reality of Everything’
CreateSpace Independent Publishing, 9/4/2018

18. “On Math, Physics, and Metaphysics”
CreateSpace Independent Publishing, 10/1/2018

19. “On Quantum Mechanics”
CreateSpace Independent Publishing, 10/15/2018

20. “On Epistemology and Ontology”
CreateSpace Independent Publishing, 10/21/2018

21. ”Toward a Metaphysics of Mass and Motion”
CreateSpace Independent Publishing, 10/29/2018

Compilations:

A. “The Universal Model of Our Sinister Universe: The First Ten Books”
CreateSpace Independent Publishing, 7/2/2017

B. “The Canons of the Sinister Universe:
The Last Four Books on the Universal Model of Our World”
CreateSpace Independent Publishing, 11/5/2017

C. “The Return of the Sinister Universe: The Immaculate Collection”
CreateSpace Independent Publishing, 9/4/2018

D. “The Battle for the Sinister Universe: The Heuristics”
CreateSpace Independent Publishing, 10/29/2018

22. “Postmodern Physics”
Kindle Direct Publishing, 10/28/2020

23. “Mechanism and Energetics”
Kindle Direct Publishing, 9/30/2021

Resources

Quantum Field Theory
Claude Itzykson, Jean-Bernard Zuber

Atomic and Quantum Physics
H. Haken, H.C. Wolf

Modern Elementary Particle Physics
Gordon Kane

Classical Dynamics of Particles and Systems
Jerry B. Marion

Foundations of Electromagnetic Theory
John R. Reitz, Frederick J. Milford, Robert W. Christy

Quantum Physics
Rolf G. Winter

Gauge Theories in Particle Physics
I. J. R. Aitchison and A. J. G. Hey

Quarks and Leptons: An Introductory Course in Modern Particle Physics
Francis Halzen, Alan D. Martin

Quantum Field Theory
F. Mandl, G. Shaw

Theoretical Mechanics of Particles and Continua
Alexander L. Fetter, John Dirk Walecka

The Theory of Spinors
Elie Cartan

Elementary Modern Physics
Richard T. Weidner, Robert L. Sells

Quantum Mechanics
Claude Cohen-Tannoudji, Bernard Diu, Franck Laloe

Philosophers

We have been hard on modern philosophers; accusing them of allowing and even aiding and abetting the physicists to run riot; riding roughshod over rhyme and reason.

In the book “Philosophical Aspects of Modern Science” (1932), C.E.M. Joad makes short shrift (in the most pleasant and politest way possible, of course) of the philosophical pretensions of Sir Arthur Eddington and Sir James Jeans. They had each proposed that Everything was Consciousness or Mind-Stuff. Were they properly chastised? Did anyone care? I don't know. The philosopher George Santayana was equally disdainful of these gentlemen’s ideas

More recently, I’ve seen some modest potshots from some philosophers at the metaphysical posturings of physicists, but they seem wan and weary. Who can blame them? We need mutual cooperation, not one more ‘culture war’. Plus, it's tiring and tedious to correct people.

A few more books:

“The Philosophy of Science”, edited by David Papineau
Oxford University Press, 1996

“Philosophy and Mystification: A Reflection on Nonsense and Clarity”, Guy Robinson
Fordham University Press, 2003

“The Investigation of the Physical World”, G. Toraldo Di Francia
Cambridge University Press, 1981

“Philosophy and Spacetime Physics”, Lawrence Sklar
University of California Press, 1985

Shop your local used book stores!

sarva asti

www.ingramcontent.com/pod-product-compliance
Ingram Content Group UK Ltd.
Pitfield, Milton Keynes, MK11 3LW, UK
UKHW061828190726
13853UKWH00009B/2499

9 798408 668045

ISBN 9798408668045